Momentum Understood
as Energy Strings and Molecular Motion

by

Mark Fennell

© 2015

Other Books by Mark Fennell

New Physics: Updating Physics with New Concepts
1. Introduction to Gravity Strings
2. New Physics of Motion: Primary Concepts

Electromagnetic Energy and Photons
1. Fundamental Properties of Electromagnetic Energy
2. Creation and Emission of Electromagnetic Energy
3. Photons in Motion

Electrical Power
1. Introduction to Electrical Power
2. Science of Batteries (in progress)
3. Hydropower Explained Simply
4. Wind Power Explained Simply
5. Solar Power Technologies Explained Simply
6. Coal Power Technologies Explained Simply
7. Nuclear Power Technologies Explained Simply
8. Natural Gas and Other Hydrocarbon Technologies Explained Simply
9. Transmission of Electrical Power Explained Simply
10. Utility Operations and Grid Systems Explained Simply
11. Efficiency and Safety of Electrical Power (in progress)

Websites and Contact Information
- Facebook.com/mark.fennell.758
- http://markfennell.blogspot.com
- YouTube: All Things Energy

About This Book

In this book we will look at motion, momentum, and friction in completely new ways. Forget what you learned in your physics class; here you will find updated and more accurate explanations of these processes.

Furthermore, instead of the traditional mathematics and abstract terms, we will discuss physical entities – the real particles and the real energy strings – which make the observed motion and momentum happen.

Also notice that these explanations are written clearly. The language is simple and easy to read.

Furthermore, there are many new concepts here. This book is part of the New Physics, where I bring physical science to a completely new level of understanding. In particular, major new concepts presented in this book include: the cause of motion for objects, the meaning of momentum, the physical entity of friction, and a more precise understanding of the speed of light.

The main topics discussed in this book will be the following:

- The Process of Motion using Atoms and Energy Strings
- Momentum explained through Energy Strings
- Self-Propelled Objects
- Energy versus Speed
- Energy Transfer Processes
- Stationary, Faster, and Slower Objects
- Sudden Stops and Turns, with Results on Passengers
- Energy Flows in Multiple Directions
- Observable Motion and Observable Momentum
- Friction Understood as Physical Entity
- Coefficient of Friction
- Friction and Momentum in the Same Situation
- Speed of Light: Constant Energy versus Constant Speed

After reading this book you will understand motion, momentum, and friction much more accurately. You will be able to apply these concepts, intuitively, to any situation involving motion and momentum.

Mark Fennell

Copyright and Fair Use

Table of Contents
for
Momentum as Energy Strings and Molecular Motion

Part A:

Preface Material

Introduction

When you are driving in a car and slam on the brakes, all the books and packages fly forward. Why is this? Physicists will call it "inertia" and "momentum". However, explaining this process as "inertia" does not give the complete understanding. A more accurate understanding would be that the molecules are continuing to travel at the same speed though the vehicle has stopped. This is the real reason.

Therefore we will take a new look at the inertia of objects in motion. Instead of discussing "inertia" and "momentum", we will discuss the following: molecules in motion, internal energy, and the degree of energy transfer.

Traditional Physics: Either Unclear or Not Accurate

When physics books or websites talk about inertia of objects, the explanations are either unclear or not accurate.

Let us first assume that the physicists do know what is going on at molecular level. We shall give them credit for such knowledge. However, if they do know these answers then they are not able to explain it to us.

On the other hand, it could also be the case that the physicists are not quite understanding the true cause of for these "inertia" situations. Most physicists do not look at the molecules or similar physical objects, they typically look at abstract "forces". Therefore it is quite possible that many of these physicists do not actually understand the true causes, at the molecular and atomic levels.

Either way, we will now remedy the situation by explaining the "inertia" and "momentum" of objects in a completely different way.

Part B:

Motion, Momentum, and Energy Strings

Momentum and Motion in Brief

For the purposes of this book, "momentum" is essentially the ability of an object to keep moving despite any activities surrounding the object. For example, when you slam on the brakes of a car, a package in the back seat will come flying forward. This is the basic concept of momentum we will use throughout the book.

We will learn in this book that the momentum is a value of energy. We will learn that the amount of momentum an object has is essentially due to the total number of energy strings contained inside the object. We will also learn that the direction of the momentum is due to the directions in which these energy strings propel the object forward.

To understand this, we must first review the understanding of "motion". Every object moves based on the motion of its molecules, and more specifically the forward motion of its protons.

Each proton is propelled by energy strings. Energy strings get inside the proton, then push the proton forward from the inside. This is the basic mechanism for motion of all particles, and particularly for protons which are the driving particles of any object.

The protons are the engines of the atom. Thus, as the protons move forward, so does the atom, and as the atoms move forward, so do the molecules. Therefore, when the majority of protons (and therefore the majority of molecules) throughout an object move forward, then the object as a whole will move forward.

In other words, every object, from a book to a planet, is self-propelled. Every object is always moving – always being self-propelled by the protons inside the object.

Consider a book. For a book to move forward, the protons must first acquire energy, and then those protons must be moving in the same direction. When enough of the protons of the book, each have enough energy, then that book will move forward.

This is the basic mechanism for motion – for any object, regardless of size, structure, or composition.

Object in Motion Tends to Stay in Motion

Once this object is in motion, it stays in motion. As Newton said "An object in Motion tends to stay in motion". This is because the object is self-propelled. The energy strings are inside the protons, and there the energy strings will remain, pushing on the protons, propelling the object as a whole forward.

Therefore, once the object (such as a book) is in motion, that object will continue to move in that direction, and at that same speed, for a long time.

Speed of Objects

Now that we have objects in motion, we will consider the speed. The speed of any object depends on the number of energy strings inside that object.

Specifically, there are energy strings located in each proton which push that proton forward. The number of energy strings then determines the speed: more energy strings inside a proton will push it faster.

Then we consider the object as a whole. There are billions of protons throughout the object – even something as simple as a book. As we add more and more energy strings to the object, these energy strings spread throughout the object, and enter into each of the protons. Therefore, with greater energy inside all these protons, the object itself will move faster.

This is how we observe an object, as a whole, to move at faster speeds.

Inertia and Momentum for
Same Object at Different Speeds

Therefore, at this stage, the concepts of inertia and momentum really mean the total internal energy of the object. Specifically, the total internal energy is the total number of energy strings, within all of the protons throughout the object.

Comparing the two books of the same size: one book with greater total energy will travel faster. This is because the total number of energy strings is greater. First this means that there are more energy strings per proton, pushing each proton faster. Second, it very likely that there are these many more energy strings in all the protons throughout the entire book. Therefore we have billions of protons, with this greater number of energy strings, pushing their protons faster. The net result is that the entire object (the book) propels itself much faster.

If no other forces act on that book, it will continue to travel at that faster speed for a long time. Similarly, if that faster book hits the wall, that book will cause more of a dent, as there is more energy which can be imparted from the book to the wall.

Thus, these examples demonstrate "inertia" and "momentum" based on molecular motion and internal energy strings.

Inertia and Momentum for Objects of Different Mass

We can also understand the concepts of inertia and momentum based on the mass of objects. It is well known that an object with greater mass will have more "inertia" and more "momentum". We can now understand this at the molecular level.

Consider two objects of very different mass – such as a book and a car. Let us get both of those objects to move at the same speed.

In any object, most of the mass is in the protons and the neutrons. Then remember that the protons are the engines of the atoms. Therefore, when an object has more mass, such as a car, it has more protons. This means that there are more "engines" of the many more atoms to be "filled" with energy.

Consequently, this will require significantly more energy strings to be distributed throughout the object in order to get the object to move. Remember that regardless of the object, we are basically getting the protons to move; and to get any proton to move at a particular speed will require a specific amount of energy strings.

The difference then is in the total number of protons throughout the object, and therefore the total number of energy strings which must be applied and distributed throughout the object.

This is why you can push the book easily with a simple press of the hand. Your hand has enough energy strings to easily be applied to the entire book, and move it forward. Yet to push the car that same speed will require several people – and often pushing their whole bodies. This is because of the amount of energy strings required to reach all the protons requires that many people give of their energy strings; and using their entire bodies helps distribute the energy strings throughout the car more easily.

Thus, this is why the difference in mass of two objects will require different amounts of energy strings applied, in order to move the two objects at the same speed.

Energy Strings Upon Impact

As a corollary: note that an object with greater number of energy strings will be able to transfer more energy upon impact. This is also an aspect of what we commonly know of as either inertia or momentum.

It does not matter if the object has more energy strings due to greater speed, or due to greater mass. The result will be the same: whenever there are more internal energy strings prior to impact, there will be more energy strings transferred after impact.

This greater number of energy strings after impact can then do several things to the impacted object. This will often propel the object forward at a fast speed, or separate the molecules from each other (resulting in cracks or causing pieces to break off).

Total Number of Energy Strings as Inertia or Momentum

Thus, again, we return to the concept of "total number of energy strings". This concept is really the central concept behind motion, speed, inertia, and momentum.

As we discussed above, when we have two objects of the same size, the one which moves faster will have more overall energy strings. This greater amount of energy strings is necessary to move the protons faster, and thus move the object as a whole much faster.

Also, as discussed above, when we have two objects of different mass, but want them to travel at the same speed, we must apply more energy strings to the object of greater mass. Remember that the number of energy strings required per proton to move each proton at a particular speed will be the same for all protons, in any object. However, the object with greater mass has many more protons, and therefore will require a greater amount of energy strings to be applied and distributed throughout that object.

In either situation, the object with the "greater momentum" or "greater inertia" will in fact have a greater total number of energy strings.

This will result in several practical effects, which are often observed. For example: it will often take longer to get a large object moving, or to move any one object faster; this is because it takes more time to apply enough energy needed, and to distribute that energy throughout the object.

Similarly, it takes longer to slow the more massive object, or a faster moving object, because all of those energy strings must be removed again.

And of course, when that object impacts another object the number of energy strings will be greater, and therefore the amount of energy transferred will be greater.

Part C:

Slowing and Stopping

Slowing and Stopping

The process of stopping is simply the reverse of getting the object to move. That is, stopping is simply the removal of all energy strings from the object.

Similarly, slowing down is simply the reverse process of speeding up. In order to get an object to move faster we must add more energy strings. Therefore, in order to get an object to move slower, we must remove some of those energy strings.

Requiring More Time to Stop

Objects with greater total number of energy strings will take longer to slow down, and longer to stop completely. This is due to the number of energy strings which must be removed.

If the object is moving very fast, then there are many energy strings per each proton. Those energy strings must be transferred, and removed from the object completely, so that all protons will slow down to the desired speed. Due to the number of energy strings involved, this can take time.

Perhaps more obvious is the object of greater mass requiring more time to slow down and stop. A classic example is a large truck. In this case there are so many protons, throughout the vehicle, that it takes a long time for the energy to transfer from all the protons to the brakes, and slow down the entire vehicle. (It is not enough for the protons near the brakes to slow down, the protons in the center of the truck must also slow down).

Therefore, the "inertia" or "momentum" is slowing down can be explained by the total number of energy strings: more energy strings total in the object will take more time (or more brakes) to slow down.

Part D:

Passengers and Packages Riding in a Car

Objects Riding on Other Objects: Overview

We are now ready to discuss momentum and inertia for objects traveling on other objects. For example, there are books and packages in the back seat of the car. There are boxes in the back of a pick-up or delivery truck. An airplane has is filled with cargo.

All of these are examples of various objects riding with a vehicle. The process of momentum can be explained using the concepts of molecular motion and energy strings described above.

Some of the most important concepts to understand are:
1. Each object propels itself, regardless of the vehicle
2. Each object has its own individual motions
3. The speed of all objects and vehicle must be the same, in order for objects to be stationary on the vehicle.
4. Any object which leaves the vehicle will move at the same speed it always had – just without the vehicle.

Individual Objects with Individual Motions

Every object is propelling itself forward, regardless of the vehicle the object sits on. Therefore each object is an individual, and its motions are individual, regardless of other objects.

It is for this reason that a box in the car can shift side to side as the car turns, or lunge forward as the car stops. The object is propelling itself independently of the vehicle. Furthermore, the object is really only held in place due to a combination of gravity and friction, which can be easily overcome.

Therefore in the situation of a box in the car, the box is independent of the car. The box is only held in place due to a combination of gravity and friction. This box is propelling itself forward, regardless of what the car is doing.

It is because of this situation that any objects sitting in the car are capable of shifting from side to side, lunging forward, or slamming backward. The object themselves are in motion, and if these objects overcome friction easily, they will continue to move straight ahead, at their constant speed, while the car changes direction and speed under the object.

Objects Riding on Other Objects will Travel at the Same Speed

Overview

Whenever we have objects in a vehicle, these objects must be traveling at the same speed as the vehicle. Remember that every object, anywhere, is propelling itself forward. Each object is an individual, and acts individually. Therefore, if we have any cargo in our vehicle, then every item must be traveling at the same speed – the same speed as our vehicle – for every object to be kept "in place" as we drive.

Must Get All Objects to Move Same Speed before Vehicle Moves

Although the objects in the vehicle are not physically connected to the vehicle, they are sitting on the chair or floor of the vehicle – in other words, physical contact with the vehicle. Therefore in order to get that vehicle to move, we need not only energy for the vehicle itself, but to all objects everywhere inside that vehicle.

Furthermore, every object in that car must be moving forward at the same speed – the speed of the vehicle itself – for the vehicle to actually move forward.

That is, the vehicle cannot physically move forward until all protons everywhere – of the vehicle and throughout all objects inside the vehicle – have enough internal energy to move their respective objects at the same speed. Then, and only then, will the vehicle as a whole – with all the cargo – be able to move forward.

It does not matter how much energy the protons of the vehicle have. Until the protons of each of the cargo objects also have enough internal energy to propel those objects at that same speed, then nothing is moving forward. Everybody moves forward at the same time, at the same speed, or nobody does.

The same is true for acceleration. In order for the vehicle as a whole to increase its speed, it is necessary for the vehicle to obtain enough energy to propel it at that speed, and then each of the objects in the vehicle must obtain enough energy to propel them at the new speed. It is only when all the energy has been added and distributed in this way will the vehicle as a whole (and every object in that vehicle) move at that faster speed.

Objects Slamming Backward
as the Vehicle Moves Forward

The situations described above for getting the vehicle with cargo to move, and to accelerate, are ideal situations. In many situations this ideal situation is what we experience.

However, there are many times where the vehicle to be in motion while the cargo remains stationary. Similarly, there are situations where the vehicle is accelerating, and yet the speed of the cargo has not been able to increase at the same rate.

These situations are what commonly experience as the vehicle lunging forward, and the cargo slamming backward. For example, a train lurches forward, and the people are slammed back into the seat. Similarly, a plane drives on the runway very fast, and the passengers are pressed back into their seats. A car speeds forward, and the box slams backward.

These are all examples of where the vehicle is moving faster than the passengers or cargo. The practical result is the vehicle seems to be getting away from the passengers, and the passengers or cargo are slammed backward. We can now understand the physical cause of this experience.

The basic explanation is that vehicle and the passengers are moving at different speeds. Therefore, because the vehicle is moving faster, the seat moves forward faster than the passenger does. Essentially, the seat is moving forward, and then hits the passenger. The passenger feels as if he is being slammed backward into the seat, but in reality it is the seat which is slamming forward into the passenger.

Here is another way to visualize the situation. Consider our train which lurches forward and slams the passengers backward into their chairs. To exaggerate what is happening, imagine that the passenger is actually floating in the air, above the seat. Remember that each object propels itself independently. Therefore the passenger and the train seat are really two separate objects, each of which can move at their own speed. Given this situation, the train starts to move. This includes all the seats of the train. Therefore all the seats move forward. But what about the passengers? If the passengers did not get enough energy strings to their bodies, they will remain in their positions. Therefore the train seats (moving forward) will slam into the passengers (which are stationary). This is what happens when the vehicle moves forward, or increases its speed, before the passengers or cargo is able to do the same.

In an ideal situation, the train would accelerate very slowly, and allow enough time for energy strings to go not only throughout the train, but through the passengers as well. This will allow the passengers to increase their speed at the same rate that the train does. In these situations, the passengers will seem "stationary" – in their positions while the train moves forward (because they are moving forward at the same speed).

However, this is not always the case. Thus, the experience of the train lurching forward, and the seats slamming into the backs of the passengers.

Also note that the visualization of the passenger floating above the seat is only a slight exaggeration. The passenger is indeed physically separate from the seat, and the passenger's motion is mostly independent of the vehicle, therefore the visual, though exaggerated, is essentially truth.

Car Turns, and Door Slams into Passenger

You have probably experienced the following. Remember when you were a passenger in a car, and this car is twisting and turning along a mountain road. During each of the turns, you may have felt yourself shift from side to side. You may have slid all the way into the door. This process can be explained by momentum.

Before we begin with the explanation, it is important to note that how we interpret the event in our minds is not always what happens on the physical level.

Why do we slide when the car turns? Simply stated: our bodies are continuing to move forward, while the car door slams into us.

Let us consider the process from the beginning. When the car is going forward, this means that all car molecules are propelling themselves forward. And for passengers to be traveling in the car, the molecules of the passengers must also be propelling themselves forward. All molecules in all objects are propelling themselves in the same direction and at the same rate.

Then the car turns. At the atomic level, this means that the energy strings pushing the car atoms are now pushing those car atoms in a new direction. Thus, the car turns.

What about the passengers? The energy strings of the humans are continuing to push these humans in that same original direction. Thus, the humans continue to travel forward, while the car beneath the humans is changing direction.

What happens next depends on the rate of the turn. If the car turns slowly, then the energy strings of the humans will change directions at approximately the same rate as the energy strings in the car. Therefore the passengers and the car will turn an approximately the same rate, and the human will shift only slightly.

However, if the car turns abruptly, or if the car was traveling very fast before the turn, then the car molecules will change direction before the human molecules will. This is what results in the feeling of "sliding across the seat", and "slamming into the door".

Again, what really happens is that the humans are continuing to propel forward. They are traveling the same direction as always. It is the car that has turned. The entire car is turning around the passengers. Therefore the seat slides *under* the passengers, and the car door *slams into* the passengers.

You will notice that the physical process is different than how our minds interpret it. We think that we are sliding across the seat to the left, yet in reality the seat is sliding under us to the right. We think we are slamming into the door on the left, yet in reality the door on the left is moving to the right, and slamming into us.

Again, when a car turns and we slide across the seat, this can be explained due to momentum – yet slightly different than perhaps we might have thought. When the car turns, the entire car around the passenger turns, yet the passengers themselves continue to move forward. Therefore the seat under the passenger moves, and the door slams into the passenger.

Notice that if our molecules are moving in the new direction, then we too will move the same direction as the car. If this occurs at the same rate at which the car molecules change direction, then the entire situation will appear "stationary". In this situation, our bodies will remain in place, in that exact spot of the seat.

Yet remember, this "stationary" situation will only occur if everything – the car molecules, the seat molecules, and our body molecules – are changing direction at the same time.

Of course passengers can be held in place due do friction between the material of the seat and the material of the person's clothing. However, this amount of friction is usually secondary to the need for human and car seat to be traveling in the same direction, and at the same speed.

Part E:

Energy Transfer and Changes in Speed

Energy Transferred from Vehicle to Objects

This leads us to the next concept: transferring energy strings from the vehicles to the objects.

How does a passenger or box obtain the energy to move at the same speed as the vehicle? The only way for this to occur is for the energy strings to transfer from the vehicle to these passengers and cargo.

Remember how it is that a vehicle moves. Energy strings are applied at one end of the vehicle – continuously, for a period of time. These energy strings are passed along, from atom to atom, from proton to proton, until all the protons throughout the vehicle have enough energy to propel the entire vehicle forward. This is how the vehicle has enough energy to move forward.

Of course, to move any object requires the same process. Yet for objects riding on a vehicle, we must go through a second stage. The energy strings must not only proceed throughout atoms of the entire vehicle, but then must also enter the objects riding on the vehicle. These energy strings must continue, being passed along throughout each object, until each object in that vehicle has enough energy to move those objects forward.

This is the process required for passengers and cargo to obtain enough energy to move themselves forward, and move themselves at the same speed as the vehicle.

You will notice that the vehicle has the engine, the power source. Therefore, the vehicle will naturally receive energy strings prior to any cargo on the vehicle. It is for this reason that it is often observed that the vehicle will move forward, or increase its speed, thought he passengers and cargo are not at that speed yet.

You will also notice that the process requires physical contact. The passenger must be sitting on the seat, and the box must be sitting on the floor. This is necessary for the energy transfer to occur.

Specifically, the atoms of the seat will bump into the atoms of the passenger, which will transfer energy strings to the passenger. And the passenger must be continuously sitting, such that additional energy strings can be transferred from the vehicle to the body of the passenger.

Then enough time must pass, while the energy strings distribute evenly throughout the passenger's body. Eventually the passenger's body will travel forward – propel itself forward – at the same speed as the vehicle.

This is how the passenger will receive enough energy from the vehicle, to eventually move forward at the same speed as that vehicle.

Of course a similar process occurs for any type of cargo. Energy strings are transferred from the atoms of the vehicle to the atoms of the cargo. When enough energy strings are transferred, and distributed throughout the cargo, that object will propel itself forward at the same speed as the vehicle.

Sudden Stops of the Vehicle
and Objects Slamming Forward

Consider the car which stops suddenly, and the packages in the back seat fly forward. Why is this? The simple answer is that the packages are moving independently of the car. Therefore when the car stops, the packages continue moving.

Remember that every object and every passenger inside a car is moving independently of the car. Each object has its own atoms, each with energy strings. These energy strings push the protons forward, and therefore propel the object as a whole forward.

The energy strings remain inside those protons, and propel that object forward, regardless what the car is doing. Therefore, the car is stopped, yet all of the objects inside the car will continue to move forward.

These objects were always self-propelled. They were always moving forward on their own. We just didn't notice it until the car stopped.

Thus again, we can understand momentum in terms of molecular motion, and inertia in terms of energy strings.

Gradual Slowing versus Sudden Stops

This leads us naturally to the preference of gradual slowing rather than sudden stops.

When we slow down a vehicle, we should think not only of the vehicle itself, but also all the passengers and cargo riding inside the vehicle. We now understand that every object and person inside that vehicle is traveling at the same speed as that vehicle.

Everything is moving at the same speed, and thus travels "together", and keeps their positions, though each item is really traveling independently. Therefore, we will want to keep everything at the same speed as we slow down the vehicle.

This is the difference between gradual slowing and sudden stops. If we slow the vehicle gradually, then the individual objects on that vehicle also have time to slow down. Yet if we slow the vehicle drastically, then the objects do not have enough time to slow down. This will result in the objects continuing to fly forward as described above.

Revisiting the Process of Slowing Down
(and the Time Required)

What exactly is the process of slowing down? To slow an object down requires that the object lose its energy strings. This is very much the reverse process of getting the object to move.

Remember that to get an object to move, such as a car, the energy strings must be applied from an external source, then these energy strings pass from atom to atom, until all the atoms throughout the object have sufficient energy to move the entire object forward.

Therefore the process of slowing down and stopping is the same process in reverse. Thus, to slow down the car we first apply the brakes. This absorbs the energy strings of the nearby atoms. Now we have a region of vehicle atoms with less energy – yet surrounding this region is the rest of the vehicle which has more energy. This creates a series of energy transfers from higher energy to atoms to lower energy atoms – and again through the brakes.

This process continues, until eventually all the atoms of the vehicle have lost enough energy for the entire car to slow down.

You will notice that this may require several applications of the brakes. Each time you apply the brakes, more of the energy strings of the car are absorbed.

If you are wondering what happens to the energy in the brakes, know that this energy is usually transferred to the surrounding air molecules. This is why the air around the brakes is always warm.

Of course if you press on your brakes firmly, without letting up, then you can stop faster. But the energy will not have a chance to transfer from the brakes into the air. This excessive energy will tear apart the material in the brakes. Thus, this is another good reason to tap on your brakes rather than holding them down continuously.

You will also notice that this process takes time. The energy strings must be transferred from the atoms of the car, to the atoms of the brakes, and into the air. This process naturally takes time. And this process takes longer if the vehicle is larger – there are more atoms of the vehicle, some of which are several feet from the brakes, which must transfer energy through the vehicle to the brakes. This is why larger vehicles will take more time to slow down.

Or, viewed from another way: remember what we said about mass and momentum. The larger vehicle has many more protons, and therefore a greater total number of energy strings. These many energy strings, spread throughout the vehicle, must be transferred through the other atoms, and to the brakes.

Therefore, the number of energy strings in the larger vehicle which must be transferred to the brakes, and the size of the vehicle (number of adjacent atoms which the energy strings must pass through) necessarily requires that the process of slowing down a larger vehicle will take more time.

Again we see that momentum is in the number of energy strings throughout the object. When we reduce the speed of the object, such as a car, we need to transfer those energy strings from their locations throughout the vehicle to the brakes and the air beyond. The amount of time this takes will depend on the number of energy strings. This is one aspect of physics we commonly refer to as "momentum".

Slowing Down Objects Inside a Vehicle

A similar process is required to slow down objects which are riding inside a vehicle. Remember that if we don't slow down these objects, then they will come flying forward when the vehicle itself stops. Therefore all passengers and cargo inside the vehicle must slow down – ideally at the same rate as the vehicle – so that everything inside the car stays in place.

The process is the same as slowing the car down. All energy strings of the protons in the passengers must transfer through the bodies, through the seats, through the car, and out through the brakes.

You can see how this process takes time. It takes time for these energy strings to transfer through all those protons, all those atoms, to finally emerge out through the brakes.

This is also very similar to a large truck, without cargo. As described above, in both cases there are many atoms which energy strings must pass through. There will also be numerous energy strings which must be transferred (due to the totally number of protons).

Again, this process of slowing takes time simply because of the number of atoms – which is both the number of energy strings, and the number of atoms the energy strings must transfer through on the way out.

This is also why it is very common for the vehicle to slow down before the passengers and cargo are able to. We can get often get the wheels and lower parts of the car to slow down quickly, yet the atoms in the main regions of the car, and the passengers, are still traveling at the higher speeds. This will result in the passengers and cargo continuing to travel forward, while the car itself has stopped.

In order to prevent this, we must slow the car gradually. This will allow all passengers and cargo sufficient time to transfer their energy strings outward, and for everything in the car to slow down at the same rate as the car.

Part F:

Stationary Objects

Stationary Objects: Main Concepts

Stationary Object Propels at Same Rate as Object Underneath

Using all this knowledge, we can now explain the exact physical concepts of "stationary" objects.

The reality is that there is no such thing as an absolutely stationary object. An object is only stationary with respect to other objects. More specifically: an object is seen as stationary only when its forward speed is exactly the same speed as the object underneath.

You are stationary to the chair because your body is propelling itself forward at the same speed as the chair. The book is stationary in the seat of the car because the book is propelling itself at the same speed as the car.

Every object which we view as stationary is simply traveling the same speed as the objects underneath. As long as there are sufficient energy strings in each object to propel each object at the same speed, then all of these objects will appear stationary to each other.

Stationary Objects on Earth are Traveling Same Speed as Earth

We should also note that most of the objects which are stationary on the Earth are in fact traveling the same speed as the Earth. Thus: every stationary rock, table, and person is in fact traveling at the same speed as the Earth. Just as we can be passengers in the car, we are indeed passengers of the Earth. Only by traveling at the same speed as our vehicle (the Earth) will our bodies have any sense of being "stationary".

This also means that in order to move the object in an observable way, we must add enough energy so that the object travels faster than the Earth.

Objects Slower than Earth are Stationary if Held by Gravity

On the other hand, objects which travel slower than Earth can also be stationary, due to gravity. Normally these objects would go flying backwards; or more precisely, the Earth would travel forward underneath the objects, leaving the object existing over a previous region of the Earth. However, there is a gravitational energy involved. The gravitational energy holds these slower objects and the Earth together. Thus, objects can be propelling themselves at a slower rate than the Earth, yet still be stationary with the Earth, because the gravity strings hold those objects to the Earth.

Therefore, a stationary object on the Earth itself will be stationary either because a) it is traveling the same speed as the Earth, or b) the gravitational energy will hold these lower energy objects to the Earth.

These concepts will be discussed in future publications.

Energy Versus Speed in Stationary Objects

Overview

However, this is not to say these two stationary objects have the same total energy. Larger objects require more energy to travel that same speed; therefore in most cases the two objects in a "stationary" situation will have different amounts of energy.

In general, there is a difference between the energy of objects and their speeds. This difference can result in several distinct concepts. One of the distinct differences is the fact that two objects can travel at the same speed, but have different energies.

More specifically: when we have two objects of different mass, and want them to travel at the same speed, the object with greater mass will require greater total energy than the object with lesser mass. This of course is related to the concepts discussed earlier in momentum.

Energy per Proton versus Total Energy of Object

First we must note that there is a difference between the energy per proton, and the total energy of the object. The amount of energy per proton will be the same, for any proton, to get that proton to travel at a particular speed.

However, the total energy in the object will depend not only on the speed that we want the protons to travel, but the number of protons in the object. It is for this reason that the total energy required to get the object with greater mass to move at a particular speed will be much more.

Energy per Proton to Reach a Particular Speed

The amount of energy required to propel a proton at a particular speed will be the same for any proton.

Remember how protons move forward: Energy strings enter the proton, push on the interior walls of the proton, and then propel that proton forward.

Furthermore, remember that to get a proton to move at a faster speed, we must add more energy strings. Having more energy strings push on the interior walls of the proton will push the proton with greater energy, and thus push the proton faster.

Therefore, if we want to get a proton to propel itself at a particular speed, then we must have a specific amount of energy inside the proton. This can be approximately understood as the number of energy strings in the proton. (Exact understanding involves not just number of strings, but also thickness and lengths). Therefore, this number of energy strings will essentially be the same for all protons, regardless where they are located.

It is for this reason that the internal energy required, per proton, for a proton to reach a specific speed, will be the same for every proton.

Total Energy in the Object to Reach a Particular Speed

Now we can see why the more massive object will require more energy to reach a particular speed. We can also see that the explanation is essentially the same as our earlier discussions of momentum due to mass.

An object with more mass has more protons, and we just learned that each proton requires the same amount of energy per proton to reach a particular speed. Therefore, because of these numerous protons in the object, more energy is required overall to get each proton throughout the object to move itself at the same desired speed.

It is for this reason that the more massive object will require a total amount of energy which is much greater than the less massive object.

General Conclusion

Therefore, objects which move at the same speed, if they are of different masses, will not have the same amount of internal energy. The more massive object will always contain more energy than the less massive object, for each to travel at the same speed.

For example, when we have a package on the seat of the car we have two objects of very different mass. The package has very little mass, not as many protons, and therefore will need only a modest amount of energy strings overall to propel that package at the desired speed. In contrast, the car has significant mass, with many more protons, and therefore requires many more energy strings overall to propel the entire car forward at that same desired speed.

Unless the two objects are identical in mass, one object must have more internal energy than the other if both objects are to travel at the same speed.

Part G:

Energy Versus Speed:

Absolute versus Relative Values

Energy Versus Speed: Additional Significant Differences

Introduction

There is another difference between the energy of an object and its speed: the energy is an absolute value, whereas the speed is measured relative to a secondary object.

This difference will become extremely important when we discuss the physical realities of the speed of light, and the flaws within Einstein's General Relativity.

Energy is Absolute Value

The amount of energy contained within an object is an absolute value. For any given object, moving at a particular speed, we will find that there is a specific amount of energy.

It does not matter where that object is traveling, or what is used as a reference point. The energy which propels the object forward is contained within that object. Therefore, the amount of energy for that object, at that moment, is an absolute value.

Speed is Measured Relative to a Second Object

The value of the "speed" of the object differs from the value of the energy, because speed can only be measured relative to a second object.

Every object in the universe is in motion. However, we need something to declare as "stationary" from which to measure the speed of our primary object. Therefore when we measure the speed of any object, we first pick another object (such as the Earth or the stars), and declare that this second object has a speed of "0.00 meters/second". Technically there is no such object, but we must do that in order to measure the "speed" of anything.

Changing Speeds With the Same Energy

This also means that the speed of an object will be measured at a different value when we use a different "stationary" object. Without adding any energy, and only by changing our reference object, the measured "speed" can be of different values. Just remember that when we do this, the energy has not changed; the amount of energy is absolute. We have only changed to a different object as our "0" speed.

Examples of Energy versus Speed

We can now use some examples of energy versus speed. Although these concepts can be applied in many situations, these concepts will become very important when we discuss the speed of light and general relativity.

For example, a car travels down the highway at 70 mph. This speed value is taken with the Earth as a stationary reference point. In reality, the earth is traveling through space very fast, but we call it stationary for convenience, and thus measure the car traveling along the road at 70 mph.

Now we have a passenger in the car. The passenger is also moving at 70 mph. We explained above how the passenger must be moving at the same speed as the car in order to remain in his seat. Thus, the passenger is traveling the same speed as the car, at 70 mph. Again, this is measured against the "stationary" Earth as our zero speed.

Yet inside the car, the passenger is not moving at all. If the car is our reference point, and we declare the car as "stationary", then the passenger is not moving at all. Remember, no energy has been changed, only our reference point for what object is "stationary".

Now let us add a second car. This car is moving at 50 mph. This speed value is again measured against the "stationary" Earth. Thus, with respect to a "stationary" Earth, we have one car traveling 70 mph, and another car moving 50 mph.

Yet suppose we change our reference point – to the second car. If we declare the second car as "stationary", then the first car is only traveling 20 mph. You know this yourself when you are on the highway looking at the cars pass by in the other lane.

And again, for all of these examples, remember that we have not changed energy one bit. There has been no increase or decrease of energy anywhere. We have only picked a different object as our reference point.

Every object is self-propelled and is traveling forward. We just simply pick an object, declare it to be "stationary", and measure the speed of all other objects relative to that declared stationary object.

Speed of Light and General Relativity

Therefore you can see that a primary difference between energy and speed is that energy is an absolute value (inherent to the object), whereas speed can only be relative value (measured against some other object).

This difference will become significant when we discuss the speed of light, and discuss General Relativity, in future publications. These publications will discuss all aspects of the light speed in great detail. At this time I will provide a preview, with some of the important concepts to understand.

1. It is said that the speed of light is constant for any observer. This is not necessarily true. What is true is that the *internal energy* of the *photon core* is constant, regardless of the *speed of the emitting source.* This statement is more accurate.

2. The internal energy of the photon core is constant. Like the car in the example above, the photon has a particular energy, and that energy is constant regardless of emission source and regardless of the speed of the observer.

3. Furthermore, the internal energy of each photon core is the same. Regardless of the frequency of the light, every photon core has the same amount of internal energy. This is another use of the word "constant" with respect to electromagnetic energy.

4. Due to the construction of the photon core, the amount of internal energy never changes. The amount of internal energy never increases or decreases. This is yet a third appropriate use of the word "constant" for the speed of light.

5. The *speed* of the photon core, however, is not technically constant. The speed must be measured according to a declared stationary object. However, the photon travels so fast that any reference point is stationary in comparison. Therefore the speed of light is "essentially constant" though not "absolutely constant"

Expanded Discussion and Analogies

More specifically: The speed of the photon core, as with the speed of the car at "70 mph", must be measured with respect to a stationary object. Technically, the speed measured would differ depending on what we declare as our stationary reference object. However, the speed of the photon is so fast, compared to any planet or star, that for all practical purposes, those planets and stars are truly stationary.

As an analogy, consider an advanced military plane, traveling at thousands of miles per hour. Compare that speed of the plane to a turtle. Then compare that speed to snail. Does it really matter whether we choose the turtle or the snail for our "stationary" object? Even if both are moving at their top speeds, the measured speed of the plane in each case will essentially be the same. That is the way it is for measuring the speed of light. It does not matter what we compare the speed of the photon to, the photon travels so fast that any reference point – even planets moving at thousands of miles per hour, seem essentially stationary by comparison.

In other words, the speed of the photon will vary depending on our reference object. (As is the case with the car speed measured 70 mph relative to Earth, but only 20 mph relative to another moving car). Yet everything else is so incredibly slow compared to the photon, that for all practical purposes the speed of light, no matter how we measure it, is "constant".

Energy is Absolutely Constant, Speed is Essentially Constant

Therefore, regarding the energy and speed of light we can say the following:
   a. The internal energy of the photon is the same, for all photons,
   b. The internal energy of the photon remains constant as it travels through space; and
   c. The photon travels so fast compared to all other objects that the photon's speed will be measured very close to the same value, regardless of how fast the reference object is moving.

Thus: the internal energy for any photon is absolutely constant, while the measured value for the speed is approximately the same regardless of how or where we measure it.

These concepts will be further developed in other publications.

Additional Notes Regarding Energy versus Speed

The concepts of energy and speed are often confused, even by some of the most eminent scientists. Therefore it is worth the time to clarify the important differences.

The various distinctions between energy and speed can be used to apply to any situation in physics involving motion. If you understand the subtleties, you will become a much greater master at all things related to energy and speed.

Understanding these distinctions now will greatly assist you in understanding not only the aspects of motion, but also electromagnetic energy, the theory of relativity, and may other advanced topics. These topics will be developed in other books.

Part H:

Launching Objects from a Moving Vehicle

Launching Objects from a Vehicle

Now that we understand the speed of passengers and cargo riding on a vehicle, we can understand the mechanics of objects thrown from a vehicle.

For example, when an airplane drops a bomb, that bomb is going forward at the same speed as the plane. This is because when the bomb was on that plane, there were sufficient energy strings to propel that bomb exactly at the same speed of the plane.

Thus the plane and the bomb were traveling together, independently yet co-existing, at the same speeds. When the bomb was let go, the bomb had those same energy strings, and thus propelled itself forward at the same speed.

This bomb would continue to fly forward at that speed – except for the gravity. Therefore the gravity pulls the bomb toward the earth. Yet the entire time the bomb is being pulled by gravity, that bomb is also propelling itself forward, due to its internal energy strings, at the exact same speed when it was riding on the plane.

Throwing a Baseball from a Train

In another example, we have a baseball thrown from a train. This is a classic physics example used frequently in texts. (This will also become important when I discuss certain aspects of electromagnetic energy in future publications).

Imagine a person standing on a train, and throwing a baseball same direction as the train is going. The speed of that baseball will mostly be the same speed of the train.

We now understand why. The energy strings of the train travel from the engine, to the train cars, up to the person standing on the train, through his arms, and into the baseball. Thus, if the guy is standing there long enough, his entire body, and the baseball he is holding, will travel at exactly the same speed as the train.

Therefore, when the man throws the baseball, the baseball will essentially be traveling the same speed as it was when the guy was holding it…and that would be the same speed as the train.

Of course the speed of the baseball can be somewhat greater depending on how much personal energy the man applies to the baseball. Yet the same process occurs: energy strings are transferred. In this case, energy strings are transferred from the man's arm and hand into the baseball. These energy strings are added to the energy strings already there. Thus, the baseball now propels itself that much faster.

Remember this important concept: we have added to the total energy of the baseball, and all energy goes into propelling the baseball in one direction. Therefore the baseball will be faster than due to the energy of the train alone. (This will become important later when we discuss electromagnetic energy and Red Shift).

We can also throw the baseball in the opposite direction, and the speed of the baseball will be less. We can again understand why.

This time we have two different flows of energy strings. Each flow of energy strings is opposite of the other.

The first energy flow is from the train. As discussed many times above, the energy strings from the vehicle transfer to the object, which allows the object to travel at the same speed as the vehicle. Thus the baseball is traveling the same speed as the train, regardless if anyone is holding it or if the ball is just sitting there.

Yet in addition we add energy strings pushing in the opposite direction. The man applies his body to the ball, which transfers energy strings to the ball – and in that direction. Therefore, when he throws the ball toward the back of the train he is sending energy strings flowing through the baseball in that direction. This is our second energy flow.

Inside the baseball, we then have two sets of energy flows. We have energy strings pushing protons in two different directions. Most of the energy strings will be pushing the baseball in the same direction the train is going (just as when the ball was on the train). Yet many other energy strings will be pushing other protons in the other direction, and therefore will push the baseball in that direction as well.

Therefore, the baseball actually travels in both directions at the same time. Some portions of the baseball will push the ball in one direction, and other portions of the baseball will push the ball in the other direction.

The baseball is then propelling itself at one speed in one direction, and another speed in the other direction. The net result is that the actual speed of the baseball will be less, Specifically, the speed of the baseball will be the speed of train minus the speed of the throw.

In this example, remember that there are two ways the energy is applied, and the energy strings are flowing in two different directions. This situation will also become important when we discuss Red Shift in future publications.

Part I:

Energy Among the Protons Throughout the Object

Protons Moving Versus the Whole Object Moving

We often talk about protons moving, and yet we also talk about the entire object moving. Now is the time where we further clarify the distinctions.

In free space, just one proton with energy in the entire object will be enough to move the object forward. The object will move slowly, but it will move.

However, on Earth most objects are held in place due to friction or gravity. Therefore the protons must have enough energy to break free from the friction.

Friction is regional. Therefore different regions of the object can break free from the friction at different times – depending on when each region gets its energy.

Therefore: some individual protons can break free from the friction and move forward, while other protons are still held by the friction. Consequently, we must get all the protons to have enough energy to break from their friction before the entire object will move forward.

That is the main distinction between protons moving and the object as a whole moving.

Distribution of Energy Among the Protons

Each proton has its own energy strings, and therefore propels itself forward at a particular speed. Yet there are many protons throughout the object. Therefore it is quite common for some protons to have more energy than others. These differences in energy can be approximately grouped in regional sections of the objects.

The main difference occurs with energy gained from an outside power source (or conversely energy lost to an outside braking device). When we first apply energy some of the protons will have more energy, and some protons will have less. The protons closer to the power source will have more energy, while the protons further away from the power source will have their original energies, and therefore lower energies.

However, given enough time the energy strings of the object will distribute among the protons uniformly. When an object allowed to have same energy for a long time, the energy strings will move back and forth among the particles, and will eventually even out. Given enough time, each proton will have approximately the same energy. Thus, the energy of each proton will be approximately uniform throughout the object. (This is of course without any new energy added from the outside or energy lost to the outside).

The reverse process occurs when the object is slowing down. Remember that an object will slow down only when a braking device is applied or there is an impact with another object. In this case, the higher energy (original energy) will be further from the braking device, while the lower energy will be at the braking device.

The process will be a general flow in reverse (as described above). Yet the main point here is this: some regions of the object will be slower than others. For example, we may have slowed down the tires of the car, yet the roof of the car is still moving at the same speed. Have you ever seen a suitcase on a car go flying forward when the car stops? This difference in energy among the regions of protons is essentially the reason why.

Part J:

Friction Energy
and
Friction with Momentum

Friction and Momentum: Overview

When we discuss observable motion there is another a factor we should consider: the presence of friction.

Every object has its internal momentum – such as the passenger or package on the car seat. However, the object may be held in place due to friction. For example, when a car suddenly stops, the package will usually go flying forward (as described above). However, if the friction between the package between the package and the seat is strong enough, the package will be held in place. This may occur despite the desire of the package to keep propelling itself forward.

Thus, although the object may still have significant internal energy, and would normally go flying forward, the object is held in place due to friction between the object and the seat.

Therefore in the following sections we will look at the physical entities of friction, and how these friction entities will affect the observable momentum of the objects.

Friction and Momentum: Basic Concepts

We will first look at the basic relationships between friction and momentum. At this stage, without understanding the physical mechanisms of friction, we are still able provide the basic relationships.

1. Momentum and Friction can be considered as two separate energies, operating in two separate directions.

Momentum is based on internal energies. These are the internal energies of all the atoms in the object. As described above, these energies push the protons forward, and collectively propel the entire object forward. The total energy involved in propelling the object forward is essentially the totality of the object's momentum.

Friction is based on external energy strings (as described later). These energy strings hold the outer portions of all molecules together. Therefore these particular energy strings pull the molecules of one object toward the other object. This causes the two objects to be held together.

Therefore we have two competing energies, operating in two different ways, and in opposing directions. Internal energy causes the objects to move forward, while the friction energy causes the two objects to be held back (and held together).

2. Whether or not an object will continue to move forward (such as the package flying forward off the seat; or staying in the seat) will therefore depend on two factors:

a) the internal energy of the upper object relative to the internal energy of the lower object (such as the internal energy of the package relative to the internal energy of the car seat), and

b) the amount of friction energy relative to the amount of internal energy in the upper object (such as the amount of friction holding the package to the seat relative to the amount of internal energy of the package).

We discussed the relative internal energies throughout the sections above. There is no need to repeat. It is the amount of friction energy versus internal energy we will now examine.

3. General Relationships

In general, if the amount of friction energy is greater than the amount of internal energy of the upper object, then that upper object will be held in place. For example, when a car stops suddenly, the package will still be held to the chair despite the fact that the package continues to move faster than the seat below.

On the other hand, if the amount of internal energy is greater than the friction energy (and that internal energy is greater than the internal energy of the object below) then the object will go flying forward. For example, when the car stops suddenly, the package will fly forward off the chair. This is because a) the internal energy is greater than that of the seat below, AND b) the internal energy of the package is greater than the friction energy holding the package to the seat.

We will expand on these concepts in the sections below.

Friction as Gravity Strings

Introduction

What exactly is "friction"? Under the New Physics, we can easily understand exactly what friction is.

The physical entity of friction is a type of energy string, known as the "gravity string". This gravity string holds all particles together, and is therefore responsible for the majority of cohesive processes in science.

A full discussion of gravity strings can be found in my book "Introduction to Gravity Strings". We will review some of those concepts, and apply those concepts to friction.

The Gravity String

The gravity string is an energy string which is physically attached to the outside of every particle. Specifically, the gravity strings are physically attached to the mass spots that exist on the outer regions of the particles.

Each gravity string pulls its particle forward, much like a strong man pulls a locomotive. When you have many gravity strings pulling the particle in the same direction, the effect is similar to a team of horses pulling on a carriage.

There are gravity strings on all regions of any particle, therefore in free space the gravity strings are pulling on the particle equally in all directions. This becomes "zero" gravity, because there is no net direction of gravitational pull.

It is only when gravity strings meet and intertwine that the mutual pull between two objects will begin. In brief, the process is as follows: gravity strings of different objects find each other in space, and begin to intertwine. It is at this time that the gravity strings have "chosen" a direction. Each gravity string is actively pulling its own object, in a straight line, along the path of the other gravity string. This is the process of mutual gravitational pull, that we commonly know of as "gravity".

Friction is the Same as Gravity

You will notice that in the New Physics, friction is the same as gravity. Both are forces which cause objects to be held together, and indeed both are a result of the same types of energy strings.

We commonly think of gravity as having an effect "up and down", while friction has an effect "side to side". Yet in reality they are the same thing.

To understand this, visualize the following: take a book and wrap a series of ropes around that book. Then take the other ends of those ropes and tie them around the desk. Set the book on the desk. Now, lift the book upward. You see the ropes stretch upward. Set the book down again. This time push the book to the side along the desk. Now you will observe that the ropes stretch to the side.

These are the exact same ropes connecting the book to the desk. Only the directions we move the book – and therefore the directions we stretch the ropes – have changed. Pulling the book and stretching the ropes upward is an example of "overcoming gravity", while pushing the book and stretching the ropes to the side is an example of "overcoming friction". Same rope, just stretching in different directions.

Note that the rope in this example is used only as a visual aid. There are in fact "ropes" connecting the book and the desk already. These are the gravity strings. If we could develop a microscope of sufficient penetration we would indeed see numerous gravity strings, exactly like the ropes above, which connect the particles of the book to the particles of the desk. This is how all objects are held together.

All Two Objects Connected with Gravity Strings/Friction

Every two objects, from protons to planets, are connected using the gravity strings. The strength of the connection depends on the total energy involved in the mutual pull.

This total energy depends on two factors: the number of intertwined energy strings, and the lengths of those energy strings. (For detailed explanation see the book "Introduction to Gravity Strings").

However, in the case of "friction" the objects begin by physically touching. Therefore the differences in lengths of the gravity strings are not as important. The main difference in the various amount of "friction" between two objects is essentially the number of gravity strings of each object which have become intertwined.

In physics, scientists have measured various values for "coefficient of friction". This value is basically an indicator for how easy or difficult it is for two objects to slide across each other. This can now be understood as the number of gravity strings which are intertwined between the two objects.

For example, when you are walking in thick mud you find it difficult to walk. This is because there are numerous gravity strings connecting your boots and the mud. The gravity strings of the mud pulls the mud to your boot, and the gravity strings of your boot pull toward the mud. There are so many gravity strings involved in this mutual pull that it requires a significant of applied energy to separate the boot from the mud.

Conversely, when you slide on ice this is because there are almost no gravity strings connecting your body molecules with the ice. There are no gravity strings connecting the two, and therefore neither are being pulled together. The ice is not being pulled to your foot, and your foot is not being pulled to the ice.

Therefore whenever we discuss the amount of friction between two objects, we are really discussing the amount of friction energy connecting the two objects; and more specifically we are considering the number of intertwined gravity strings pulling each object toward the other.

Friction versus Gravity: Subtle use in Real World Applications

When two objects physically touch and are held in place, we can call this energy either "friction" or "gravity". Generally, "friction" is used to represent the resistance of pushing one object across another, while "gravity" is used to represent the resistance of an object being lifted into the air.

It should also be noted that objects have multiple gravity strings at the same time, and therefore can be connected to various objects at the same time. For example, a book is connected to the desk with gravity strings, yet the book is also connected to the ground with gravity strings. There are different sets of gravity strings involved, but the book has gravity strings in use with both the desk and the ground at the same time.

This brings us to one of the subtle differences in "friction" versus "gravity" in many real world applications. When we speak of friction energy, we are indeed talking of gravity strings, yet only of the gravity strings between objects which physically touch (such as the book and the desk). For other gravity strings, such as those connecting the book and the ground, this is more commonly considered "gravitational energy".

I bring this point up for general understanding. In this book we are focusing on Friction. Yet I want the reader to be able to understand the physical reality in any of these situations.

Friction Holding Objects in Place
Despite Sudden Stops and Turns

Introduction

Now that we understand the basic concepts of friction (as gravity strings connecting two objects) we are ready to understand friction in our discussions of momentum.

As above, we will continue to focus our discussions on packages or passengers sitting in a seat. We will look at the amount of friction relative to the amount of internal energy, to see what will happen when the car suddenly stops or turns.

Friction Holding the Object Back when Car Stops Suddenly

Note that for this discussion we will assume that the internal energy (momentum) of the package has NOT slowed down or turned at the same rate as the car. Therefore, if there were no friction, the package would continue to propel itself forward.

This means, of course, that if the package does not fly forward, the only reason would be due to sufficient friction between the package and the seat.

Imagine a dog on a leash. The dog has his own energy, and tries to run forward. However, you are holding onto the leash, which holds the dog back. Thus you can see that the dog will be held in place, despite the fact that he has his own significant energy propelling him forward.

The same situation exists with our package on the seat of the car. The package, like the dog, has its own internal energy. The package is propelling itself forward. Yet there are energy strings, much like the leash, which hold the package to seat. These are gravity strings, which provide the energy we call "friction" (as described above). Therefore, the package is held in place, despite the fact that the package is propelling itself faster than the car seat.

General Process of Friction Holding Back Objects

We can therefore see the basic process of friction on objects with momentum.

Where one object sit on top of another, such as a package on a seat, and where the upper object has greater speed than another (such as where the car stops suddenly), the higher speed object would continue to fly forward due to its internal energy, or "momentum".

However, this object (the package) may be held in place despite its greater speed due to friction. This friction energy is physically contained in gravity strings which will pull the two objects together. If the friction energy holding the two objects together (package to seat) is great enough, then like the dog on the leash, the object will be held in place, despite its momentum.

When Momentum is Greater than Friction

Overview

We can now consider the opposite case: where the momentum is greater than friction. Specifically this means that the internal energy of the object, propelling the object forward, is much greater than the friction energy (gravity strings) pulling the object back. In this case, the object will propel itself forward.

The Scenario

Again we will assume that the object has enough internal energy to propel itself forward, without considering friction. For example, we slam on the brakes, and the package on the seat flies forward. We explained above that this will because the internal energy of the package has not changed, though the car has lost some of its internal energy. Therefore the package moves faster than the car, and continues flying forward.

Now we consider friction. In the previous example we discussed the process of the friction energy holding the package to the chair – despite the momentum of the package trying to propel the package forward. In this example we consider the opposite situation: where the friction energy is not enough to hold the package in place. Therefore, the package flies forward – though perhaps not as far.

Analogy of the Dog on the Leash

We can again return to the analogy of the dog on the leash. The dog on the leash has his own energy, and he desires to run forward. Yet we are holding onto the leash, pulling back. If we are able to hold onto the leash with greater energy than the dog's energy of propelling himself forward, then the dog will remain in place, despite all of his energy.

However, if the dog has significant energy, then he may be able to break free from your hold. You may have experienced this. When a dog has significant energy, he is able to tear away from the leash and pull forward. He will be able to run free, as he desires.

The situation with the package on the seat is very similar. If the package has much more internal energy than the energy of the gravity strings, then the package will propel itself away from the gravity strings – and hence away from the seat. The package can then fly forward; very similar to the dog who has propelled himself away from the leash.

Amount of Momentum and Separating from Friction

Overview

You may also notice at this point that the amount of momentum an object has will be a factor in whether or not the object will move forward.

Momentum as Number of Internal Energy Strings

Remember that "momentum" is really the total number of energy strings inside all the atoms of the object. A greater total number of energy strings is equivalent to greater momentum; while a smaller total number of energy strings is equivalent to less momentum.

Also remember that this total number of energy strings can be arrived at by either the speed or the mass. Regarding the speed and its momentum: for an object to go faster we must have more energy strings inside the protons. This will propel each of those protons at a faster speed, and hence the entire object at a faster speed. Therefore, any object which travels faster will have greater momentum, due to the greater number of energy strings.

Regarding the mass and momentum: remember that when an object has greater mass, this means the object has more protons and neutrons. Further remember that protons and neutrons are the engines of the atoms. Therefore if we have an object with greater mass, and we want that object to move faster, then we must apply the same additional amount of energy strings to all the protons and neutrons throughout the object.

Consequently, an object with greater mass will naturally have greater momentum, due to the number of energy strings required to reach all of the protons throughout the object.

Thus, any object with greater mass or greater speed will naturally have more energy strings as the object propels itself forward. This total internal energy becomes the basis for "momentum" of the object.

Amount of Momentum and Friction in Observable Motion

We can now apply this understanding of "momentum" to the situations involving "friction". We can better understand the relative energies involved.

In our situations of momentum versus friction, we have competing energies. The internal energies of the atoms in the object are propelling the object forward. (Such as our package). At the same time, the gravity strings on the outside of the atoms are pulling the object backward, toward other objects. (Such as the gravity strings pulling package and car seat together).

The net result will depend on which sets of energies are greater. If the total internal energy is greater, than the object will propel itself away from the gravity strings, and become a free entity. Conversely, if the total friction energy is greater, then these energy strings will hold the object back.

Number of Friction Strings / Coefficient of Friction
is also a Factor

Overview

You may also notice that the number of friction strings is also a factor. Just as more internal energy strings is what creates greater momentum, more gravity strings intertwined is what creates a stronger amount of friction.

Therefore, when two objects are connected with gravity strings, the amount of friction energy holding the two objects in place will be determined by the number of gravity strings which are intertwined.

Coefficient of Friction

Furthermore, and as described elsewhere, this physical reality is essentially what creates the "coefficient of friction" between two specific materials.

This value is essentially the number of gravity strings intertwined between two materials, and will be the same value per any area of each object touching. In other words, it does not matter how large either of the two objects are (how much material is involved). All that matters is that the object has the same material throughout.

Add Enough Internal Energy to Overcome Friction Energy

Therefore, in order to get one object to move past another object, including overcoming the friction, we must add enough internal energy per each square region of the object to break free from the gravity strings in that same region. This will allow us to push the object beyond the initial region.

How much energy is required? This depends on the number of gravity strings intertwined, which is also the coefficient of friction. If there are only a few gravity strings intertwined, then only a small amount of energy is required to be added. Conversely, if there are numerous gravity strings intertwined, then we must add significant amount of energy to the object to get that object to break free from the gravity strings.

Continuous Gravity Strings and Continuous Friction

This leads us to the next concept: that gravity strings are everywhere. This is why objects tend to stop due to friction. It is also why a plane must keep in continuous motion or it will fall to the Earth.

Every object has gravity strings extending from it. We can apply enough internal energy so that the object breaks free from those gravity strings. However, a few inches ahead, and a few feet ahead, there will be more gravity strings. Different locations of gravity strings, but the same types.

Whether the gravity strings are involved as friction (we want to push along a surface) or as gravity (we want to move upward or forward in the air), the concept is the same. There are gravity strings everywhere, and if we want to keep our object in motion (continuously) then we must continue to add energy to our object, which will allow our object to continually break free from the next set of gravity strings encountered.

Part K:

Review and Summary
of
Momentum, Motion, Friction,
and Energy Strings

Review of Momentum as
Energy Strings and Molecular Motion

In the preceding sections we discussed momentum as energy strings propelling objects forward. We can now take a look back at what we have discussed, and present some highlights.

Energy Strings and Momentum

1. Energy strings are the basis for all motion and all momentum.

2. Protons are the engines of all molecules and all objects. Energy strings enter these protons, and push these protons forward.

3. When the majority of protons propel themselves in a particular direction, then the object as a whole will propel itself forward, in that direction.

4. Momentum is essentially the total number of energy strings within the protons of the atoms. When an object has more energy strings overall, then that object will have greater momentum. Conversely, when an object has fewer energy strings overall, then that object will have less momentum.

5. Momentum is a result of speed or mass. Regarding speed, in order to get an object to move faster we must add more energy strings to the protons, in order to push those protons faster.

Regarding mass, there are more protons throughout the object, which requires more energy strings added, in order to add energy to each proton throughout the object.

Either way, an object in motion with greater mass or greater speed will require more energy strings. This means more total energy, and greater momentum.

Self-Propelled Objects

6. All objects are self-propelled, and all objects have internal energy. We may or may not observe this due to the speed of the self-propelled objects nearby.

7. In order to get a car to move forward, energy must be added not only to the molecules of the car, but to all the passengers and objects inside the car.

Every object must have enough energy to move with approximately the same speed as the car in order for the car, and its contents, to move.

8. Each object propels itself independently. Each object inside a car travels independently, based on its own internal energy strings. The speed of each object can be affected by the other objects which are touching. However, each object is essentially independent, propelling itself forward.

Stationary, Faster, and Slower Objects

9. Objects are said to be "stationary" when they propel themselves at the same speed. For example, a package is stationary to the car if that package is propelling itself the same speed as the car.

In addition, objects can also be stationary if the gravity strings hold the objects together.

10. An object or a passenger in a moving car will be "stationary" with respect to the car if the object or passenger is propelling itself at the same speed as the car.

11. If the object is propelling itself faster than the car, then the object will fly forward through the car.

12. Conversely, if the object is propelling itself slower than the car, then the object will slam backward into the seat. More precisely, the faster moving seat of the car is slamming into the slower moving object.

13. When a car turns suddenly, a passenger may feel himself sliding across the seat. In reality, the passenger is continuing to move forward in the same direction, while the seat is sliding under him.

Similarly, when a car turns suddenly the passenger may feel himself slamming into the side door. In reality, the car door is slamming into the passenger as the he continues to move forward.

Energy Transfer Processes

14. The mechanism for energy transfer is energy strings which exit one proton and enter the neighboring proton. This is the process for each proton (and hence each atom and molecule) to lose energy in the one particle, and to gain energy in the other particle.

  This process is done most effectively through the collision of protons. As one proton collides with another, energy strings from one proton will migrate from one proton to the other.

15. In order for an entire object to move, such as a car, the majority of car protons must have sufficient energy, and these protons must be propelled in the same direction.

16. The usual process for adding energy to an object (such as to a car) is as follows: Energy is usually applied from one region (where the fuel is located) to the nearby car atoms. These car atoms collide with nearby other car atoms, and thus pass on the energy strings. This process continues, until those specific energy strings are in the atoms furthest from the fuel source.

  At the same time, we continue adding fuel (adding more energy strings). Therefore, more energy strings are added, and passed along. This is done continually, until the vast majority of protons in the car have enough energy to propel the entire car forward.

  We continue to add fuel (energy strings) to the system to increase the speed of the car. Specifically, the energy strings are added to the first set of atoms, then passed along throughout the atoms of the car. Eventually the majority of protons will have the same amount of additional energy strings. This new amount of energy strings will propel each proton at a faster speed, and collectively move the entire car at a higher speed.

17. Energy is transferred to objects and passengers in the car in exactly the same process. As we add fuel, and allow enough time, the new energy strings will be transferred from proton to proton throughout the car, and into the protons in each object and each passenger.

Eventually each passenger and each package will propel itself at the same speed as the car. When the passenger propels himself at the same speed as the car, he feels "stationary" with respect to the car, though both are moving forward at the same speed.

18. The process for decreasing speed is the same general process but in reverse. Instead of adding fuel to the car, we add brakes. The brakes have lower energy, and therefore energy strings are transferred from the car atoms to brake atoms.

This causes a successive chain of energy string transfers. All energy strings are channeled to the lower energy particles, and eventually outward to the brakes, and into the surrounding air molecules.

In this way, all molecules of objects and people inside a car will transfer their energy through the car, and outward into the brakes and the air.

Observable Motion and Observable Momentum

19. The difference in the speed between two objects is what produces "observable motion". For example, when a passenger propels himself faster than the car, we will observe the passenger moving.

In reality, the passenger was always moving. It is simply that he was previously moving at the same rate as the car, and therefore appeared stationary. When the car suddenly stops and the passenger keeps moving at the same speed, we can then see him moving.

20. The rate of energy transfer from one object to another is what will most often cause observable momentum.

It takes time for energy to be transferred throughout an object (such as a car), and to each object in the car. If energy strings are added slowly, then all objects will increase in energy at the same rate as the car, and all passengers, objects and the car will increase their speed at the same rate.

However, if the car is given a sudden burst of energy, there may not be time for energy to be transferred to passengers. In this case, the back of the car will slam into the passengers. (Passengers feel they are being slammed into the seat; the result is the same, but the cause is slightly different).

Similarly, it takes time for energy to transfer as a decrease in energy. If the driver slams on his brakes, there may not be enough time for the energy strings of the passengers to transfer outward through the car. Therefore in this case, the passengers are continuing to propel themselves at the same speed, while the car is moving much slower. This results in passengers flying forward inside the car.

21. If an object, such as a package, is propelling itself at a faster speed than the object below, such as the car seat, then the package will fly forward.
     However, this can object can be restrained or held in place due to friction.

Friction and Momentum

22. Friction is the same as Gravity. Both processes use the same energy strings, and operate with the same mechanism.

23. The physical entity of friction energy is the "gravity string". The gravity string is attached to the outside of the particle, and pulls the particle forward.
     When the gravity strings from two particles meet and intertwine, these gravity strings pull their respective particles together. This begins the mutual gravitational process. This process is the mechanism which brings all two masses together.
     All aspects of gravity and friction are based on the mutual pull of these gravity strings.

24. The effect of friction energy on the momentum of objects can be compared to a leash on a dog. The dog has his own internal energy, and propels himself forward. In the same way, an object has internal energy, and propels itself forward.
     However, if the dog is attached to a leash, and a person holds that leash, he can keeps the dog in place, despite the internal energy of the dog. In the same way, the friction strings pull on the outside of the particles, thus holding the particles in place, despite the internal energy of those particles.

25. The "coefficient of friction" is a measurement of how easy or difficult it is to slide two specific materials across each other. This value is essentially determined by the number of gravity strings which are intertwined.
     When two materials have more gravity strings intertwined, the materials are held more tightly together via those gravity strings, and therefore more energy is required to slide the materials past each other.

Conversely, when two materials have fewer gravity strings intertwined, the materials are not held together as tightly, and therefore not as much energy is required to slide the materials past each other.

26. When we have a situation involving both friction and momentum, then net result will depend on which type of energy is greater.

Assuming that the object (such as a package in a car) has enough energy to travel faster than the object below, and the only other factor to consider is friction, then:

If the friction energy is greater, then the object will be held in place. (Such as a package held into the seat)

If the internal energy is greater, then the object will fly forward. (Such as the package flying forward).

Energy Flow in Multiple Directions

27. Energy strings can push the object in multiple directions at the same time.

For example, in a baseball energy strings can push different regions of the baseball in different directions. We can throw the baseball such that the energy strings in one region of the ball push forward, while energy strings in another region push to the right. This can result in curve balls and other unusual motions.

Thus, in region by region of an object, it can be the case where the particles, per each region, are propelling the object in different directions. This can result in the object moving in a complex series of motions.

28. There can also be multiple energy flows at the same time. We discussed flow of energy above: energy strings passing from proton to proton. In addition, if the energy strings are added at different locations of the object, then multiple energy flows will exist at the same time. This can produce a combination of momentum effects, and diverse internal motions, inside of the object.

These multiple energy flows can also be produced when particles collide at particular angles. This can result in the particles moving in specific directions (including the process of recoil).

Additional Details and Discussion

Additional discussion of these topics, with further details and illustrations, can be found in my following publications:

1. Introduction to Gravity Strings
2. New Physics of Motion – Primary Concepts
3. A New Conceptual Understanding of Energy Strings